YOUR KNOWLEDGE HAS VALUE

Bradley Tice

Compression and Geometric Data

GRIN Verlag

Bibliografische Information der Deutschen Nationalbibliothek:

Die Deutsche Bibliothek verzeichnet diese Publikation in der Deutschen National-
bibliografie; detaillierte bibliografische Daten sind im Internet über http://dnb.d-
nb.de/ abrufbar.

Imprint:

Copyright © 2012 GRIN Verlag GmbH
Druck und Bindung: Books on Demand GmbH, Norderstedt Germany
ISBN: 978-3-656-64527-6

This book at GRIN:

http://www.grin.com/en/e-book/199139/compression-and-geometric-data

GRIN - Your knowledge has value

Der GRIN Verlag publiziert seit 1998 wissenschaftliche Arbeiten von Studenten, Hochschullehrern und anderen Akademikern als eBook und gedrucktes Buch. Die Verlagswebsite www.grin.com ist die ideale Plattform zur Veröffentlichung von Hausarbeiten, Abschlussarbeiten, wissenschaftlichen Aufsätzen, Dissertationen und Fachbüchern.

Visit us on the internet:

http://www.grin.com/

http://www.facebook.com/grincom

http://www.twitter.com/grin_com

Compression and Geometric Data

Bradley S. Tice

Advanced Human Design, P.O. Box 3868 Turlock, California 95381

ABSTRACT

Kolmogorov Complexity defines a random binary sequential string as being less patterned than a non-random binary sequential string. Accordingly, the non-random binary sequential string will retain the information about it's original length when compressed, where as the random binary sequential string will not retain such information. In introducing a radix 2 based system to a sequential string of both random and non-random series of strings using a radix 2, or binary, based system. When a program is introduced to both random and non-random radix 2 based sequential strings that notes each similar subgroup of the sequential string as being a multiple of that specific character and affords a memory to that unit of information during compression, a sub-maximal measure of Kolmogorov Complexity results in the random radix 2 based sequential string. This differs from conventional knowledge of the random binary sequential string compression values.

PACS numbers: 89.70 Eg, 89.70 Hj, 89.75 Fb, 89.75 Kd

Traditional literature regarding compression values of a random

binary sequential string have an equal measure to length that is

not reducible from the original state[1]. Kolmogorov complexity

states that a random sequential string is less patterned than a

non-random sequential string and that information about the

original length of the non-random string will be retained after

compression [2]. Kolmogorov complexity is the result of the

development of Algorithmic Information Theory that was discovered

in the mid-1960's [3]. Algorithmic Information Theory is a sub-

group of Information Theory that was developed by Shannon in 1948

[4].

1

Recent work by the author has introduced a radix 2 based system, or a binary system, to both random and non-random sequential strings [5]. A patterned system of segments in a binary sequential string as represented by a series of 1's and 0's is rather a question of perception of subgroups within the string, rather than an innate quality of the string itself. While Algorithmic Information Theory has given a definition of patterned verses patternless in sequential strings as a measure of random verses non-random traits, the existing standard for this measure for Kolmogorov Complexity has some limits that can be redefined to form a new sub-maximal measure of Kolomogorov Complexity in sequential binary strings [6]. Traditional literature has a non-random binary sequential string as being such: [111000111000111] resulting in total character length of 15 with groups of 1's and 0's that are sub-grouped in units of threes. A random binary sequence of strings will look similar to this example: [110100111000010] resulting in a mixture of sub-groups that seem 'less patterned' than the non-random sample previously given.

Compression is the quality of a string to reduce from it's original length to a compressed value that still has the property of 'decompressing' to it's original size without the lose of the information inherent in the original state before compression.

2

This original information is the quantity of the strings

original length before compression, bit length, as measured by

the exact duplication of the 1's and 0's found in that original

sequential string. The measure of the string's randomness is

just a measure of the patterned quality found in the string.

The quality of 'memory' of the original pre-compressed state of

the binary sequential string has to do with the quantity of the

number of 1's and 0's in that string and the exact order of those

digits in the original string are the measure of the ability to

compress in the first place. Traditional literature has a non-

random binary sequential string as being able to compress, while

a random binary sequential string will not be able to compress.

But if the measure of the number and order of digits in a binary

sequence of strings is the sole factor for defining a random or

non-random trait to a binary sequential string, then it is

possible to 'reduce' a random binary sequential string by some

measure of itself in the form of sub-groups.

These sub-groups, while not being as uniform as a non-random sub-

group of a binary sequential string, will nonetheless compress

from the original state to one that has reduced the redundancy in

the string by implementing a compression in each subgroup of the

random binary sequential string. In other words, each sub-group

3

of the random binary sequential string will compress, retain the

memory of that pre-compression state, and then, when

decompressed, produce the original number and order to random

binary sequential string.

The memory aspect to the random binary sequential string is, in

effect, the retaining of the number and order of the information

found in the original pre-compression state. This can be done by

assigning a relation to the subgroup that has a quality of

reducing and then returning to the original state that can be

done with the use of simple arithmetic. By assigning each sub-

group in the random binary sequential string with a value of the

multiplication of the amount found in that sub-group, a quantity

is given that can be retained for use in reducing and expanding

to the original size of that quantity and can be represented by a

single character that represents the total number of characters

found in that sub-group.

This is the very nature of compression and duplicates the process

found in the non-random binary sequential strings. As an

example the random binary sequential string [110001001101111] can

be grouped into sub-groups as follows: {11}, {000}, {1}, {00},

{11}, {0}, and {111} with each sub-group bracketed into common

families of like digits. An expedient method to reduce this
string would be to take similar types and reduce to a single
character that represented a multiple of the exact number of
characters found in that sub-group. In this case taking the
bracketed {11} and assign a multiple of 2 to a single
character,1, and then reduced it to a single character in the
bracket that is underlined to note the placement of the
compression. The compressed random binary sequential string
would appear like this: [1000100101111] with the total character
length of 13, exhibiting the loss of two characters due to the
compression of the two similar sub-groups.

De-compression would be the removal of the underlining of each
character and the replacement of the 1's characters to each of
the sub-groups that would constitute a 100% retention of the
original character number and order to the random binary
sequential string. This makes for a new measure of Kolmogorov
Complexity in a random binary sequential string.

Summary

The use of a viable compression method for sequential binary
strings has applied aspects to transmission and storage of
geometric data. Future papers will explore practical
applications to industry regarding applied aspects of compression
to geometric data.

5

Reference

[1] S. Kotz and N.I. Johnson, Encyclopedia of Statistical
     Sciences (John Wiley & Sons, New York, 1982).

[2] abide.

[3] R.J. Solomonoff, Inf. & Cont. 7, 1-22 & 224-254 (1964), A.N.
     Kolmogorov, Pro. Inf. & Trans. 1, 1-7 (1965) and G.J.
     Chaitin, Jour. ACM 16, 145-159 (1969).

[4] C.E. Shannon, Bell Labs. Tech. Jour. 27, 379-423 and 623-656
     (1948).

[5] B.S. Tice, Aspects of Kolmogorov Complexity: The
     Physics of Information. (River Publishers, Denmark, 2009).

[6] S. Kotz and N.I. Johnson, Encyclopedia of Statistical
     Sciences (John Wiley & Sons, New York, 1982).